Monster numbers

Identify one more and one less

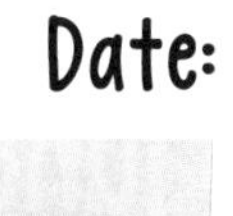

Date:

• coloured pencils

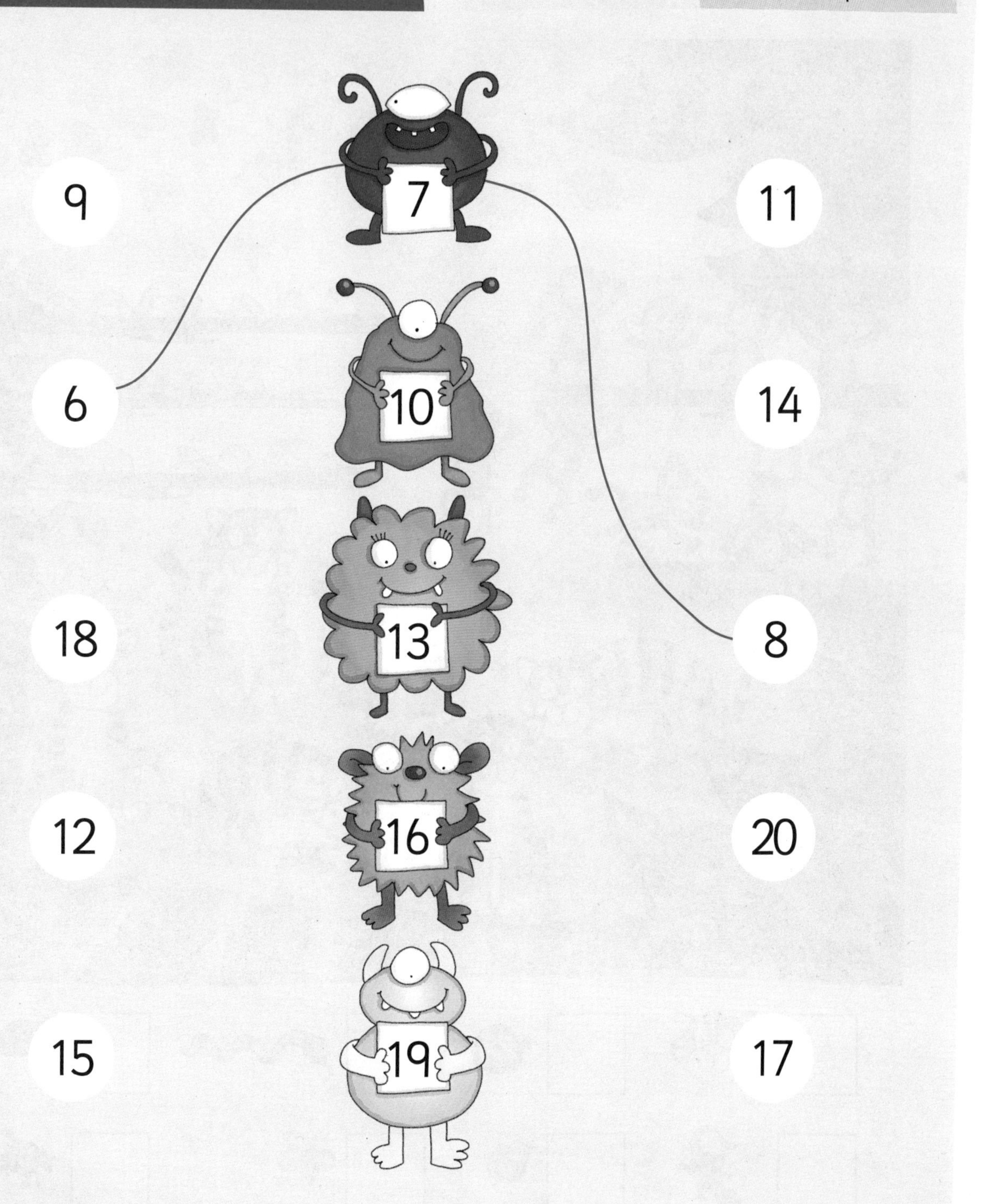

Teacher's notes

Children look at the number each monster is holding. Using the same colour as the monster, they draw a line to join it to the number that is one less and the number that is one more.

Date: ______________

Counting in the jungle

Count and write numbers to 20

Teacher's notes

Children count how many there are of each creature and write the number in the correct box.

Collins

Activity Book 1A

Date: ____________

What's my number?

Read and write numbers 0 to 20

Teacher's notes

Children identify the missing numbers and write them in the empty boxes to complete the sequence 0 to 20.

Date: ______________

Garden planting

Practise ordering 1st to 20th

You will need:
- coloured pencils

Colour the **3rd** and the **6th**.

Colour the **7th** and the **9th**.

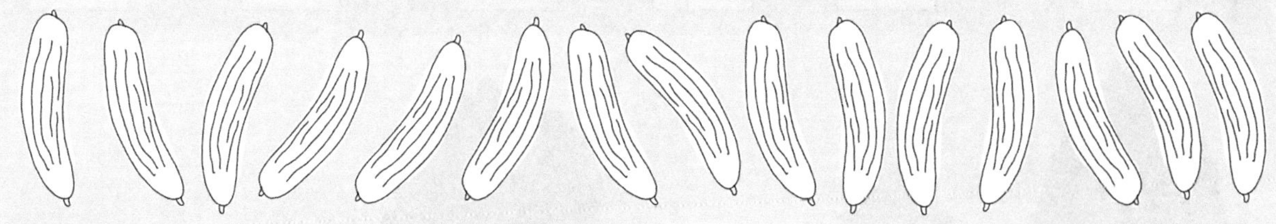

Colour the **11th** and the **14th**.

Colour the **1st**, the **17th** and the **13th**.

Colour the **12th**, the **19th** and the **last**.

Teacher's notes

Support children to follow the instructions to colour the correct flowers and vegetables.

Date: ________________

Apple tree addition

Make addition number sentences to 5 by joining groups

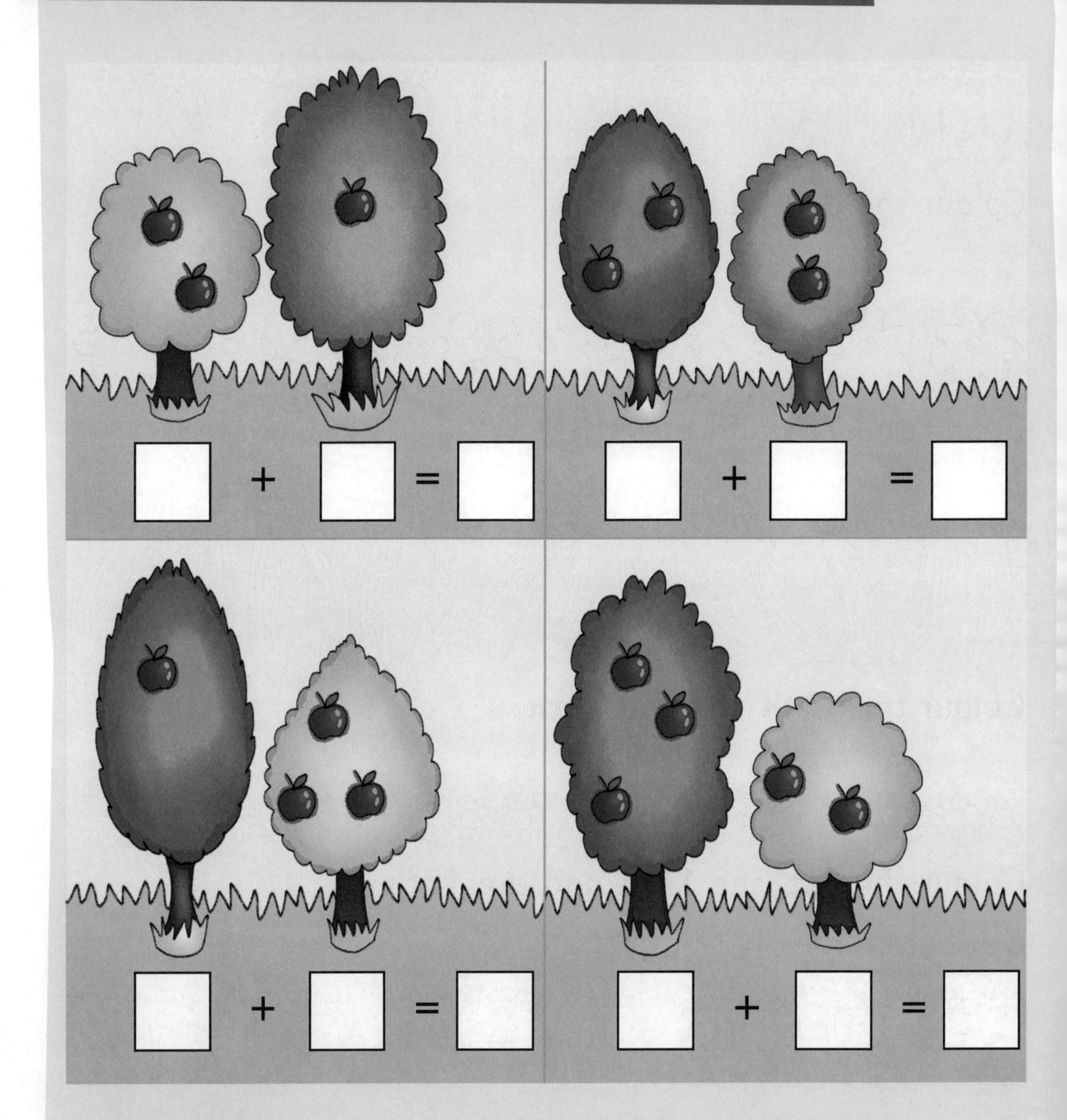

Teacher's notes

For each picture, children count the apples in the first tree and write the number in the first box. They count the apples in the second tree and write the number in the second box. They then complete the addition calculation, writing the answer in the space provided.

Date: ____________________

Adding on stepping stones

Make addition number sentences to 5 by adding on

Teacher's notes

For each picture, children write the number of the stone the character is standing on in the first box. They trace the number of jumps the character says they are going to take, and write the same number in the second box. They complete the calculation by writing the total, which is the number of the stone the character lands on, in the last box.

Date: ____________________

Take away trees

Make subtraction number sentences to 5 by taking away

Teacher's notes

For each picture, children count the apples on the tree and write the number in the first box. They cross out the number of apples to be taken away and complete the subtraction calculation by writing how many apples are left.

Date: ________________

Subtraction stepping stones

Make subtraction number sentences to 5 by taking away

Teacher's notes

For each picture, children write the number of the stone the character is standing on in the first box. They trace the number of jumps back the character says they are going to take and write the same number in the second box. They complete the calculation by writing the answer, which is the number of the stone the character lands on, in the last box.

Date: ____________________

Shape names

Know circles, triangles, squares and rectangles

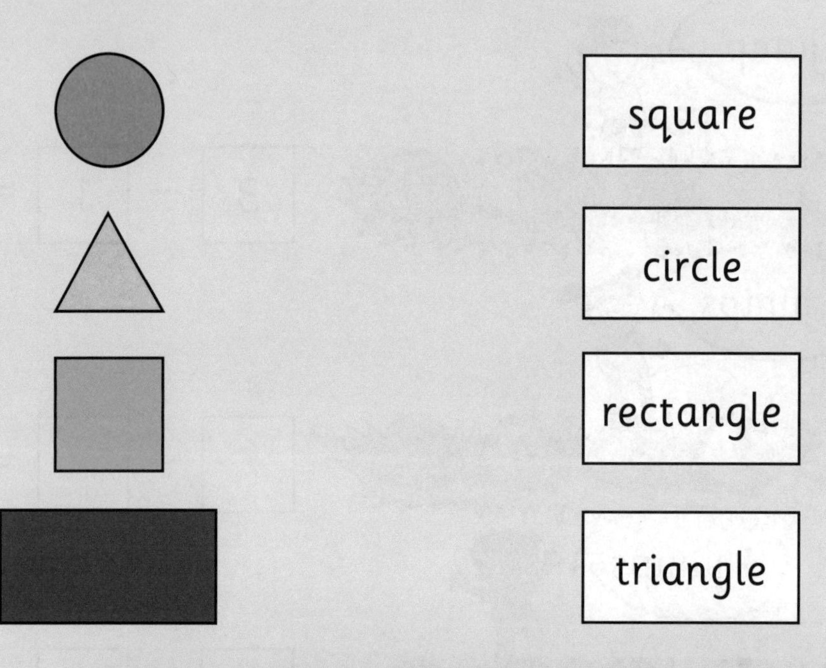

	Number of sides	Number of corners

Teacher's notes

Children draw lines to match the names to the shapes. Then they count and write how many sides and corners each shape has in the table.

Date: ___________________

Robot shapes

Spot circles, triangles, rectangles and squares

You will need:
- green, purple, yellow and blue coloured pencils

Key

Teacher's notes

Children colour each type of shape in the colour shown.

Date: ____________

Triangles

Know which shapes are triangles

You will need:
- coloured pencils

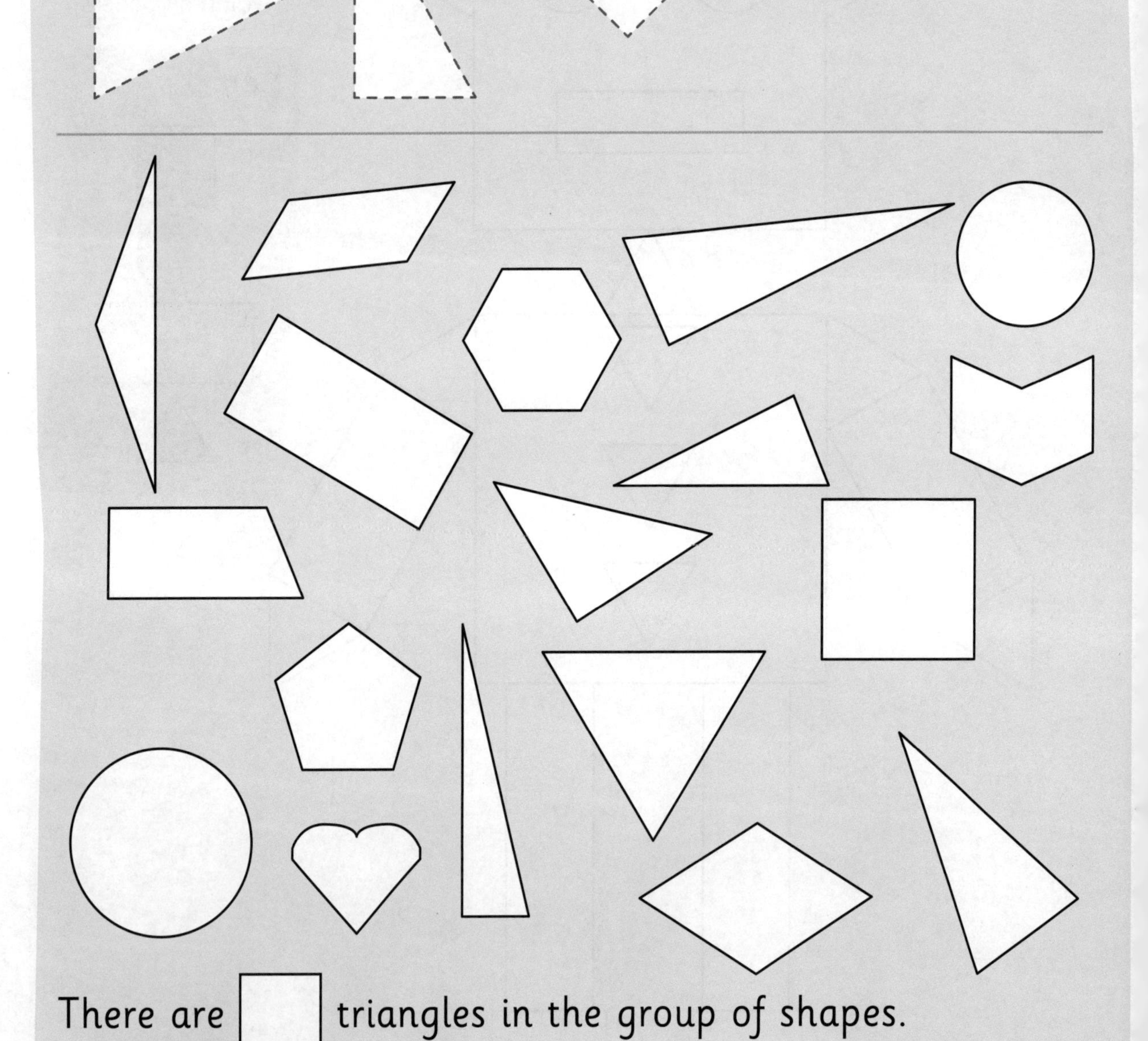

There are ☐ triangles in the group of shapes.

Teacher's notes

Children trace over the dashed lines to draw in the sides of the group of four triangles. They then colour all of the triangles in the group of 2-D shapes. Finally, they count and write how many triangles there are in the group of shapes.

Date: ______________

Rectangles and squares

Know and draw rectangles and squares

You will need:
- red and blue coloured pencils

This is a [ ].   This is a [ ].

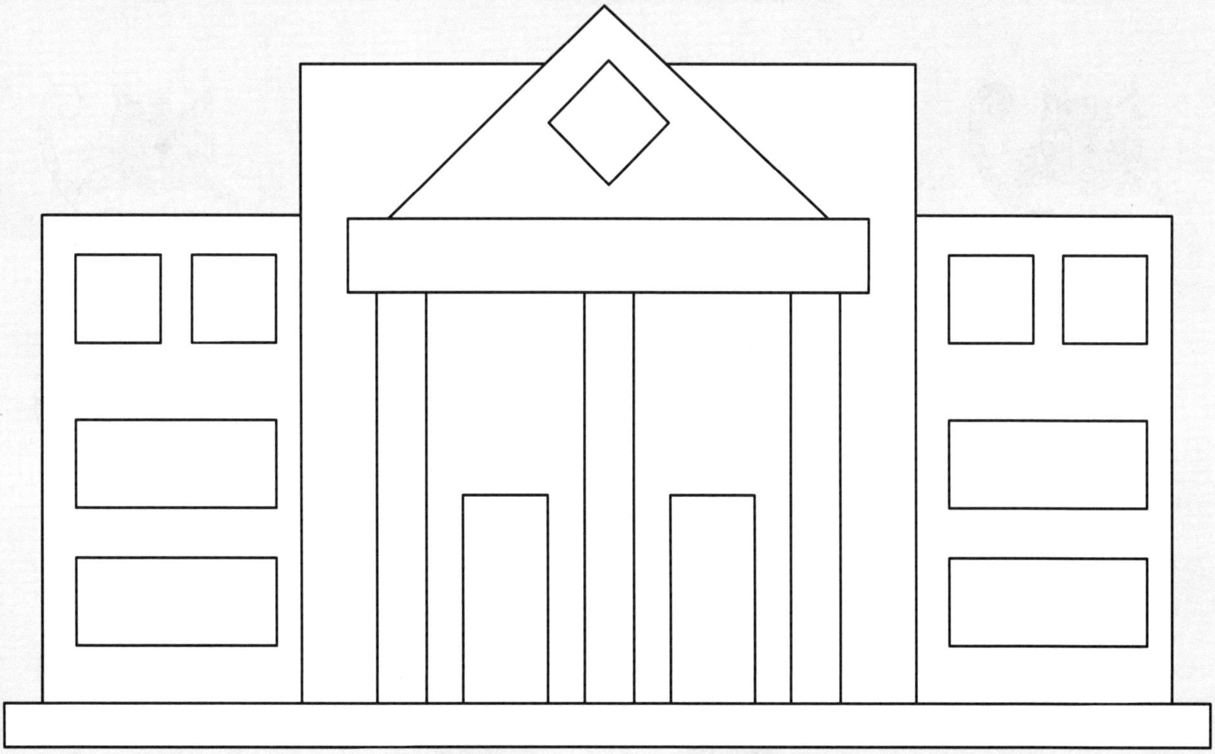

Teacher's notes

Children draw the shapes by tracing over the dashed lines and colouring them: one shape red, the other shape blue. They then complete the number sentences to name them. Finally, they colour the shapes in the picture to match.

Date: ____________________

Addition fact cats to 10

Know addition facts to 10

Teacher's notes

Children draw a line to match each addition fact cat with the wall showing the answer.

Subtraction cats to 10

Date: ______________

Know subtraction facts to 10

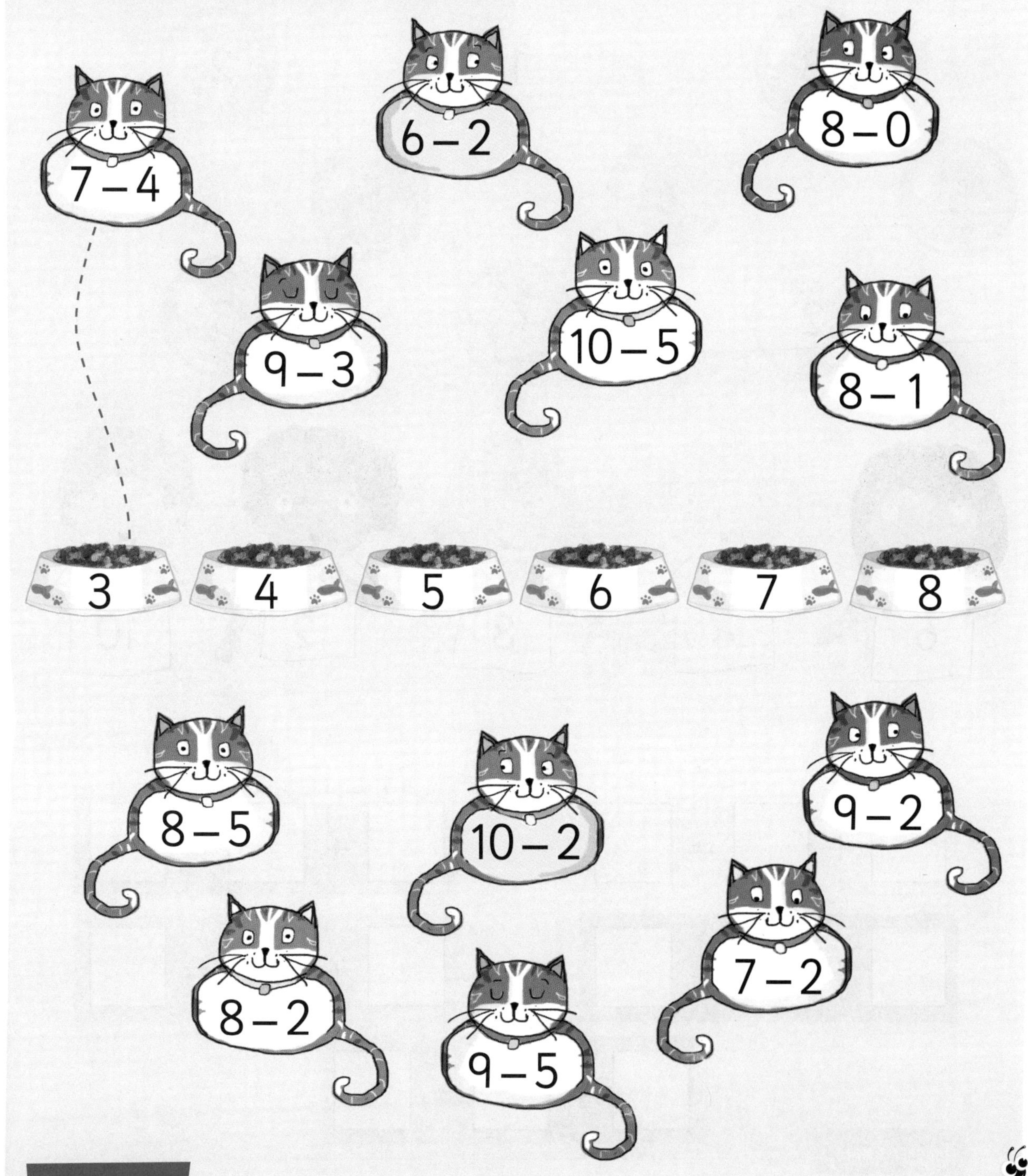

Teacher's notes

Children draw a line to match each subtraction fact cat with the food bowl showing the answer.

Date: ______________

Bubble doubles

Know addition doubles to 5 + 5

You will need:
- coloured pencils

3 1 2

5 4 5 1

2 4 3

6 4 8 2 10

☐ + ☐ = ☐ ☐ + ☐ = ☐

☐ + ☐ = ☐ ☐ + ☐ = ☐

☐ + ☐ = ☐

Teacher's notes

Children look for two bubbles containing the same number. They then look for the child wearing a t-shirt with the doubles total and colour this to match. They write the doubles calculation in the corresponding coloured boxes.

Date: ____________________

Cake calculations

Match addition and subtraction facts to 10

2 + 1 = ☐

☐ + ☐ = ☐

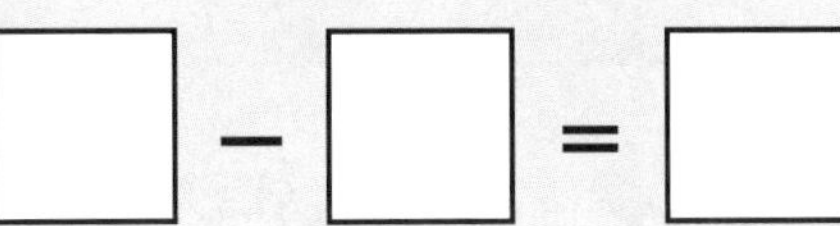

☐ + ☐ = ☐

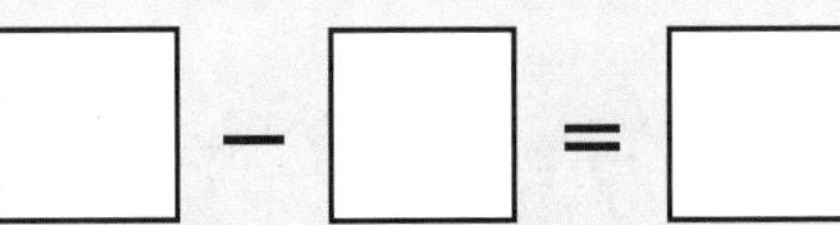

☐ + ☐ = ☐

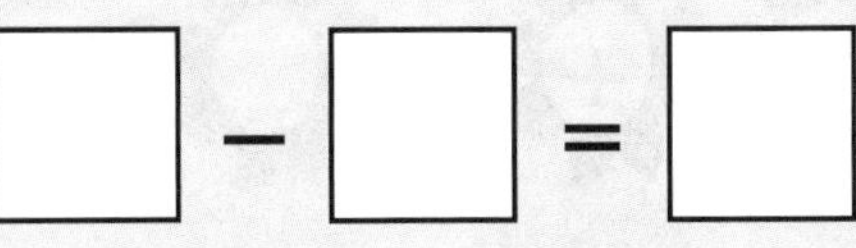

☐ + ☐ = ☐

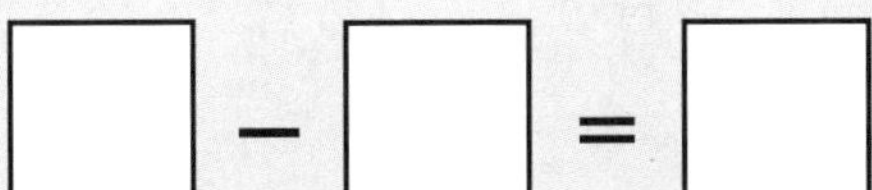

Teacher's notes

Children write an addition calculation to represent the plate of cakes on the left and a related subtraction calculation to represent the plate of cakes on the right.

Date: ______________

Flower facts to 10

Add numbers in any order

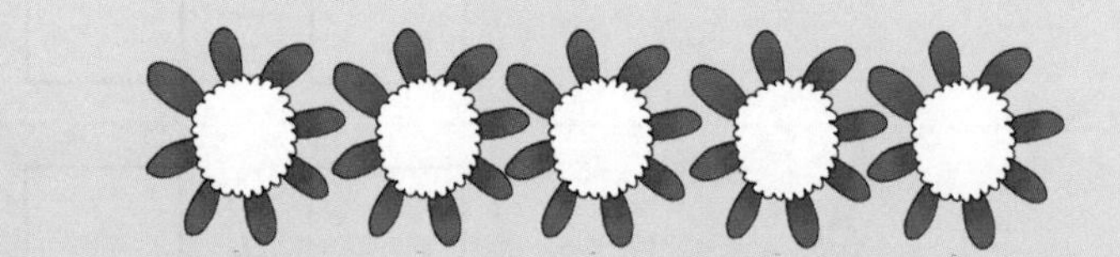

Teacher's notes

Children rewrite the addition calculations in the correct order, then complete each one by writing in the answer.

Date: ____________________

Dinosaur difference

Find the difference between two numbers

Teacher's notes

Children use the number track to find the difference between the numbers on each pair of dinosaurs. They write this as a subtraction number sentence in the spaces provided.

Date: ______________

Plant problems

Solve missing number problems

Teacher's notes

Children complete each calculation and write the missing number in the space provided.

Date: ____________

Add or subtract?

Solve addition and subtraction problems

There are 3 apples on one tree and 4 on another tree.

How many apples are there altogether?

☐ ☐ ☐ ☐ ☐

There were 8 strawberries on the plate.

Amber ate 4 of them.

How many were left?

☐ ☐ ☐ ☐ ☐

Amir made a tower of 7 bricks.

He added 3 more.

How many bricks were there altogether?

☐ ☐ ☐ ☐ ☐

There were 10 socks on the washing line.

4 blew away in the wind.

How many socks were left?

☐ ☐ ☐ ☐ ☐

Teacher's notes

Children write the addition or subtraction calculation for each problem in the spaces provided.

Date: ____________________

Which is longest?

Talk about and compare lengths

Teacher's notes

In each set of animals, children label the longest animal 'L' and the shortest 'S'.

Which is tallest?

Talk about and compare heights

Date: ______________________

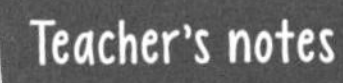

In each row of animals, children label the tallest animal 'T' and the shortest 'S'.

Measuring with hands and feet

Date: ______________________

Measure lengths, widths and heights

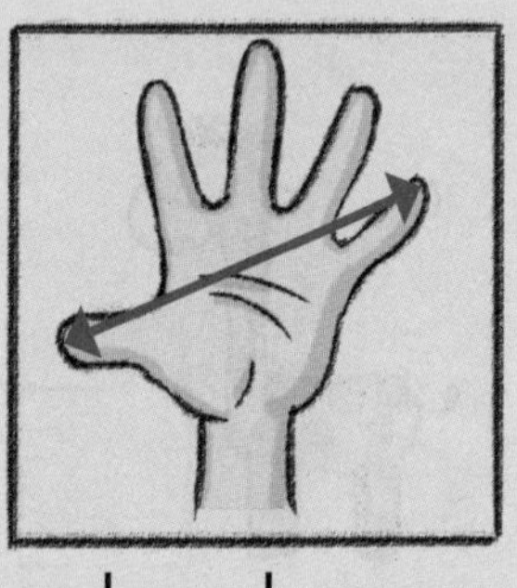

a hand span

a stride

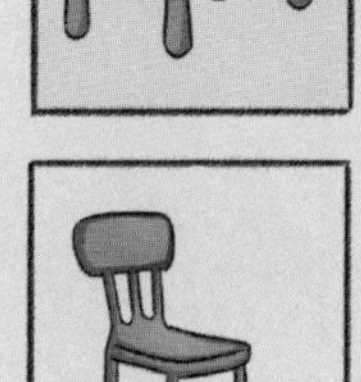

The table is ☐ spans wide.

The chair is ☐ spans tall.

The playground is ☐ strides long.

The bookcase is ☐ spans wide.

The corridor is ☐ strides long.

Teacher's notes

Ensure children have access to each of the objects and spaces shown. They measure and record each length, width or height, in hand spans or strides.

Date: ____________

Measuring length, width and height

Use a ruler to measure lengths, widths and heights

You will need:
- ruler

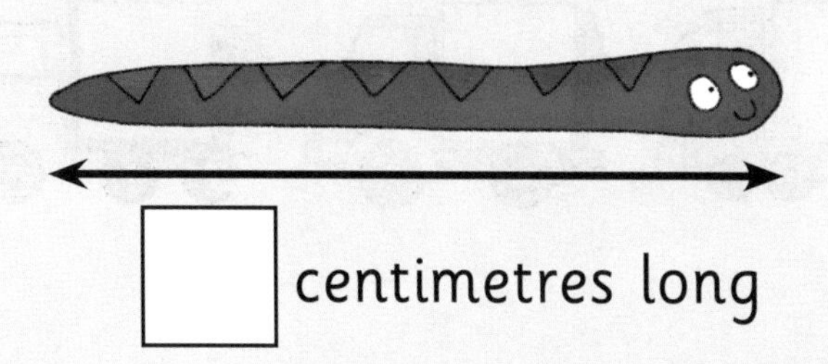

☐ centimetres long

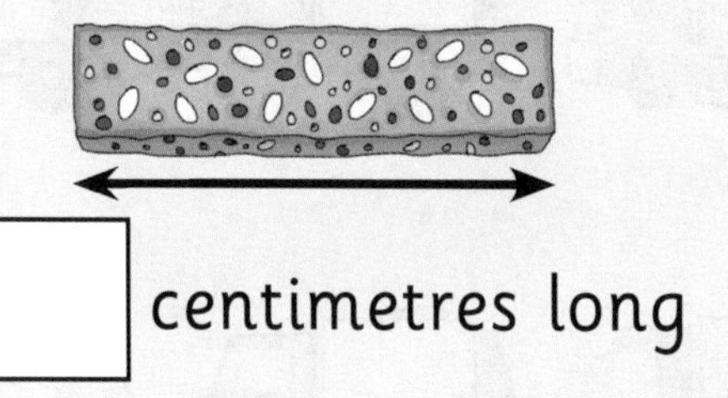

☐ centimetres long

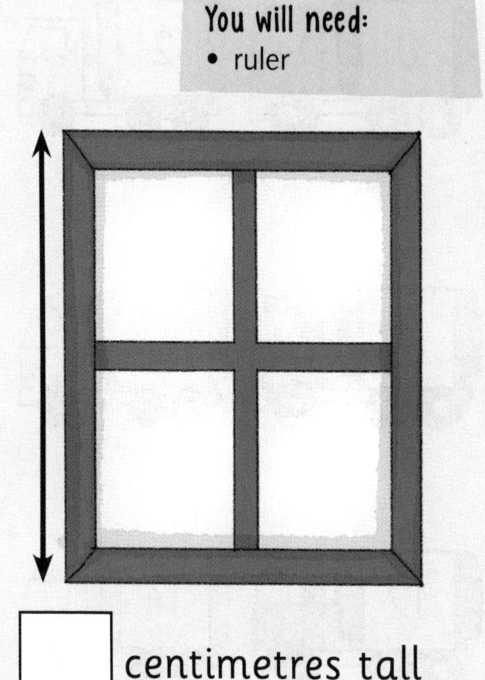

☐ centimetres tall

☐ centimetres long

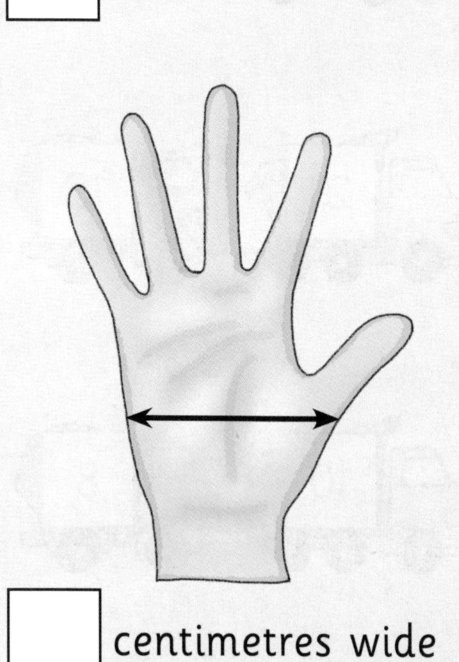

☐ centimetres wide

Teacher's notes

Children use a ruler to measure each item and record the length, width or height, in centimetres.

Date: ______________________

Truck 2s

Count in 2s

Teacher's notes

For each row, children count on or back in twos and write in the missing numbers.

Date: ______________

Fishy 5s

Count in 5s

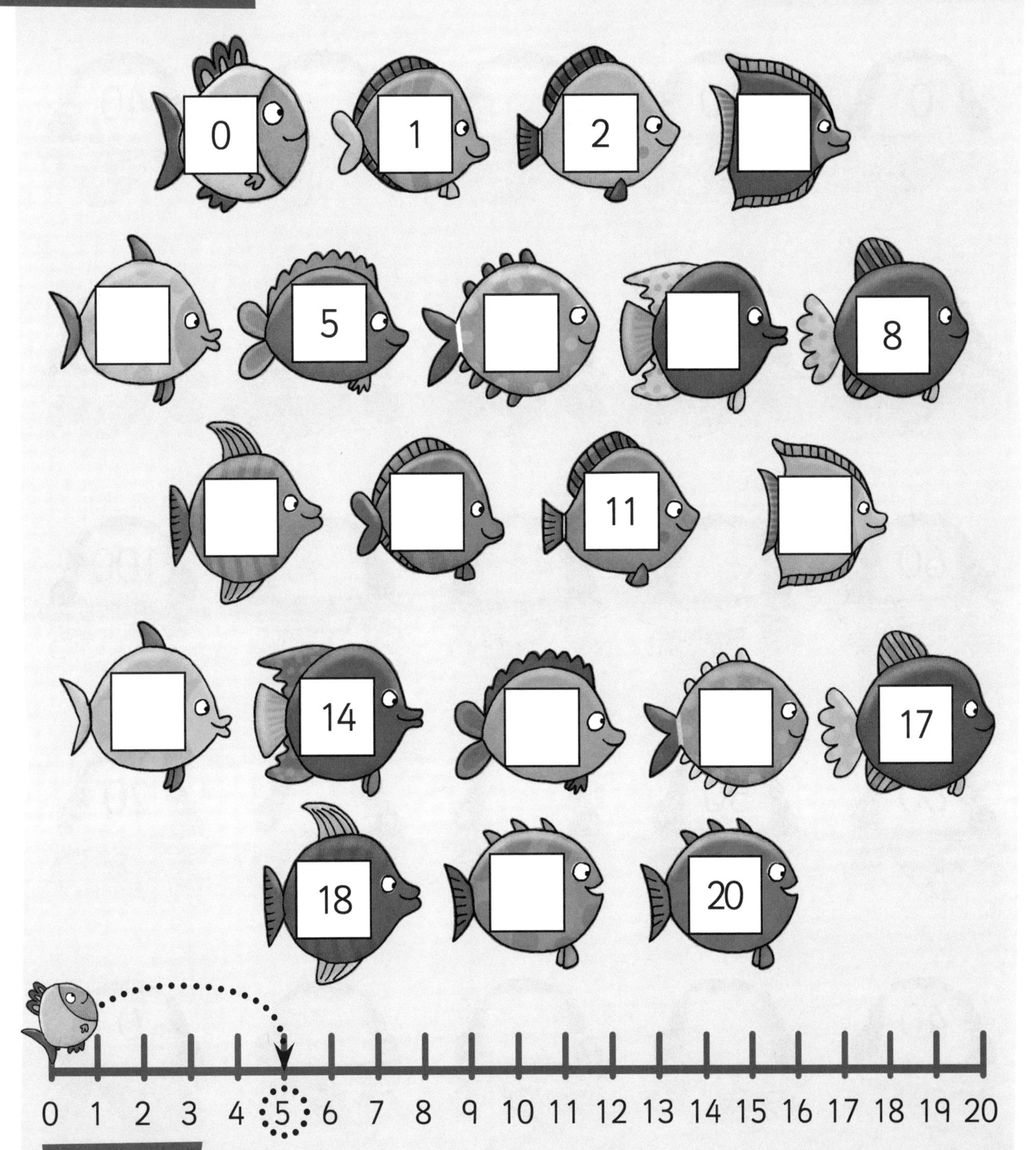

Teacher's notes

Children write in the missing numbers from 0 to 20. They then count on from 0 in steps of 5 and circle each fish that is a multiple of 5. Then on the number line, they move the fish from 0 in steps of 5, drawing a circle around each number the fish lands on.

Date: ____________________

Turtle shell 10s

Count in 10s

Teacher's notes

For each row, children count on or back in tens and write in the missing numbers.

Number patterns of 2s, 5s and 10s

Date: ___________________

Count in 2s, 5s and 10s

Teacher's notes

In each chain, children count on in twos, fives or tens starting from zero each time, and write in the missing numbers.

Date: ______________

Cars of 2

Count sets of 2

0	2	4								20

1 set of 2 makes ☐.

How many people altogether? ☐

☐ sets of ☐ makes ☐.

How many people altogether? ☐

☐ sets of ☐ makes ☐.

How many people altogether? ☐

☐ sets of ☐ makes ☐.

How many people altogether? ☐

☐ sets of ☐ makes ☐.

How many people altogether? ☐

Teacher's notes

Children complete the number track to show the multiples of 2 from 0 to 20. Then for each question, they draw two faces in each car, write how many sets of 2 there are and then how many people there are altogether.

Date: ______________

Apple 5s

Count sets of 5

0	5				25

1 set of 5 makes ☐.
There are ☐ apples altogether.

☐ sets of ☐ makes ☐.
There are ☐ apples altogether.

☐ sets of ☐ makes ☐.
There are ☐ apples altogether.

☐ sets of ☐ makes ☐.
There are ☐ apples altogether.

☐ sets of ☐ makes ☐.
There are ☐ apples altogether.

Teacher's notes

Children complete the number track to show the multiples of 5 from 0 to 25. Then for each question, they draw five apples on each plate, write how many sets of 5 there are and then how many apples there are altogether.

Date: ______________

Pen 10s

Count sets of 10

0	10				50

[ ] pack of pens make

[ ] pens altogether.

[ ] packs of pens makes

[ ] pens altogether.

[ ] packs of pens makes

[ ] pens altogether.

[ ] packs of pens makes

[ ] pens altogether.

[ ] packs of pens makes

[ ] pens altogether.

Teacher's notes

Children complete the number track to show the multiples of 10 from 0 to 50. They write how many packs of 10 pens there are in each group, then how many pens there are altogether.

Date: ____________

Sharing strawberries

Share into equal sets

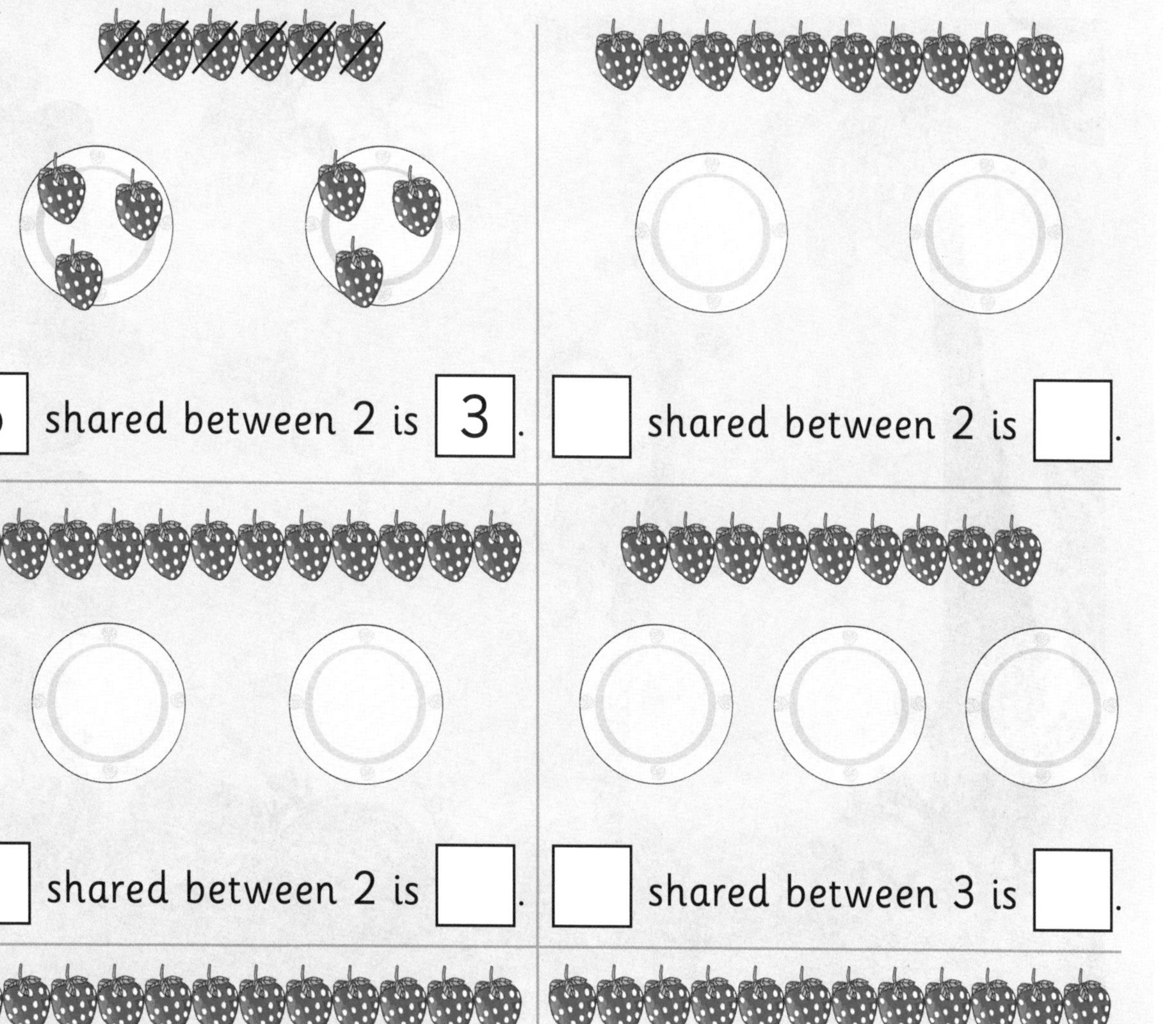

6 shared between 2 is 3.

☐ shared between 2 is ☐.

☐ shared between 2 is ☐.

☐ shared between 3 is ☐.

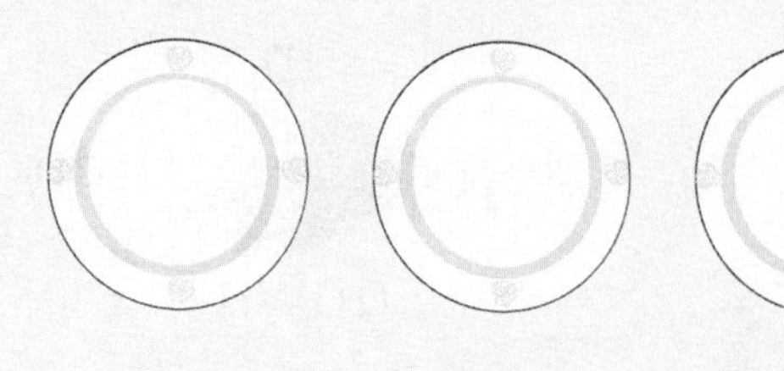

☐ shared between 4 is ☐.

☐ shared between 3 is ☐.

Teacher's notes

Children count each group of strawberries and then draw them on the plates, sharing them to make equal sets. They then complete each sentence by writing in the correct numbers.

Date: ____________________

Monkey directions

Use the words 'up', 'down', 'left' and 'right'

up left [watermelon] right down	up left [grapes] right down
up left [bananas] right down	up left [peanuts] right down

Teacher's notes

Children circle either 'up' or 'down' and either 'left' or 'right' to give the monkey directions to reach each food.

Date: ______________

Where is it?

Use position words

You will need:
- coloured pencils

What is above the ? 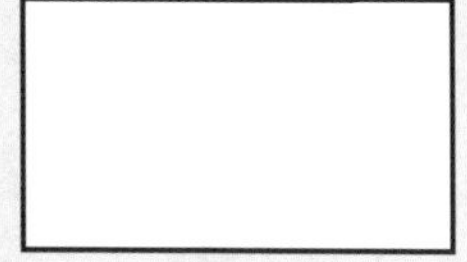

What is below the ? 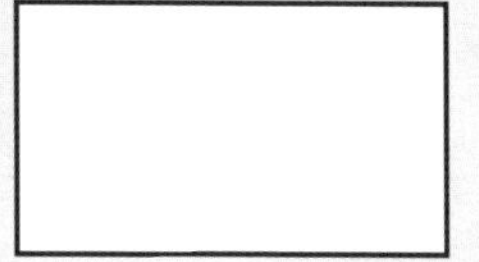

What is between the 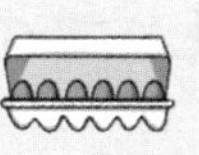 and the ? 

What is between the  and the ? 

Teacher's notes

Children label the shelves 'top', 'middle' and 'bottom'. They answer the questions by drawing the food in the position described.

Date: ___________________

Whole and half turns

Know whole and half turns

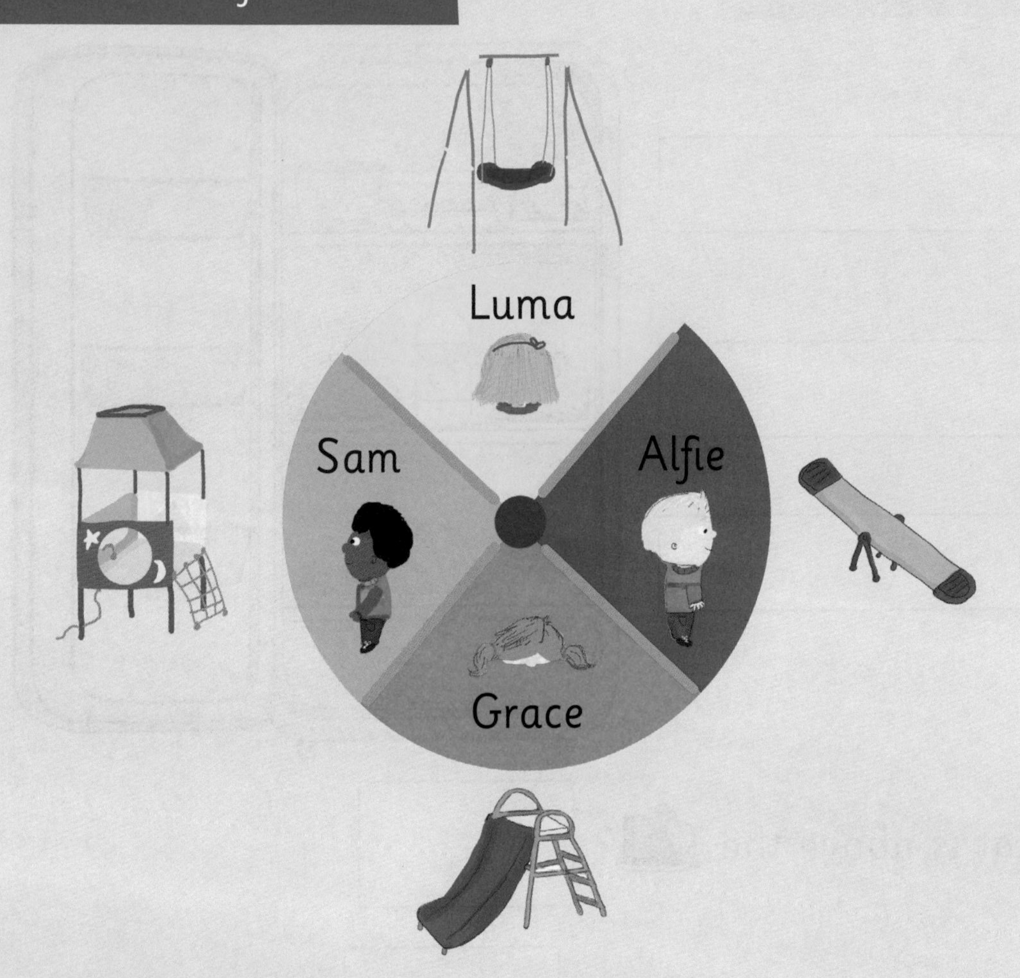

After a **whole turn** 

Luma  sees

Alfie  sees

Grace  sees

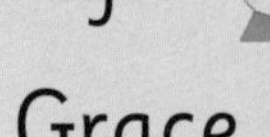

Sam  sees

After a **half turn** 

Luma sees

Alfie sees

Grace  sees

Sam  sees

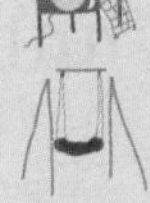

Teacher's notes

Children draw a line to show what each character would see after a whole turn and after a half turn.

Quarter and three-quarter turns

Date: ____________________

Know quarter and three-quarter turns

You will need:
- coloured pencils
- small-world figure (optional)

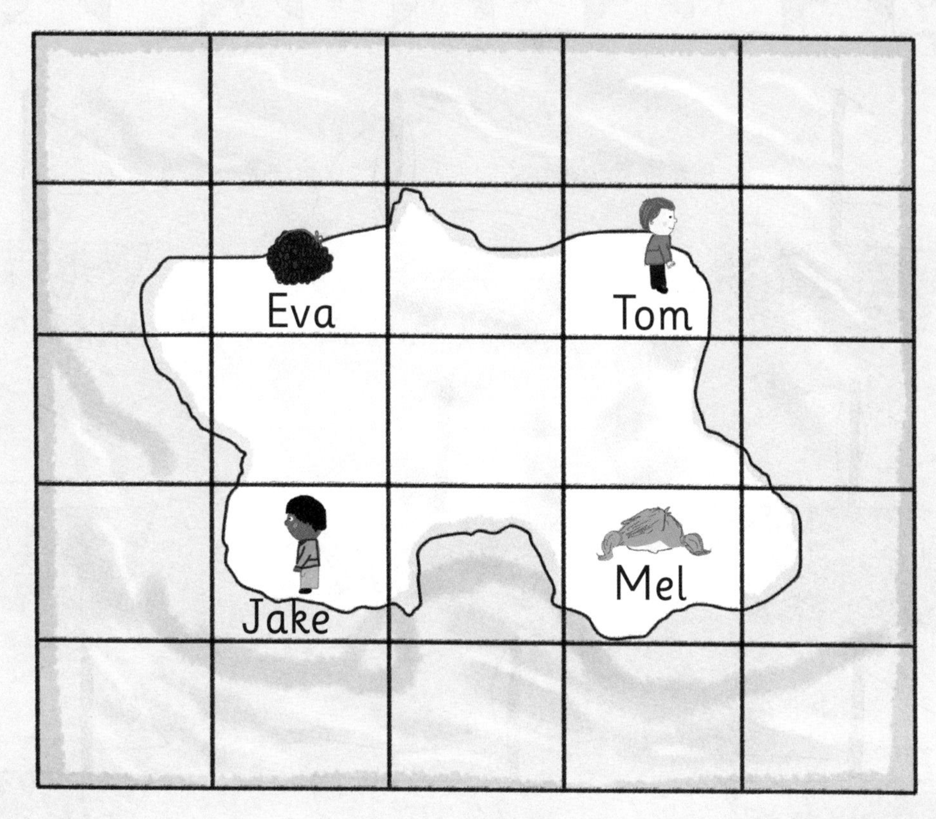

	Eva	Tom	Jake	Mel
quarter turn to right ↷				
three-quarter turn to right				

Teacher's notes

Children draw the objects the characters would see after a quarter turn and after a three-quarter turn on the map. Remind them to check which way each character is facing to begin with. They can use a small-world figure to help them work out the answers.

Date: ______________

Anchor addition

Know addition facts to 15

You will need:
- coloured pencils

Teacher's notes

Children work out the answer to the addition fact on each boat. They find the answer on an anchor and colour the boat's sails to match. There will be two sails of each colour.

Take away trains

Know subtraction facts to 15

Date: ____________________

You will need:
- coloured pencils

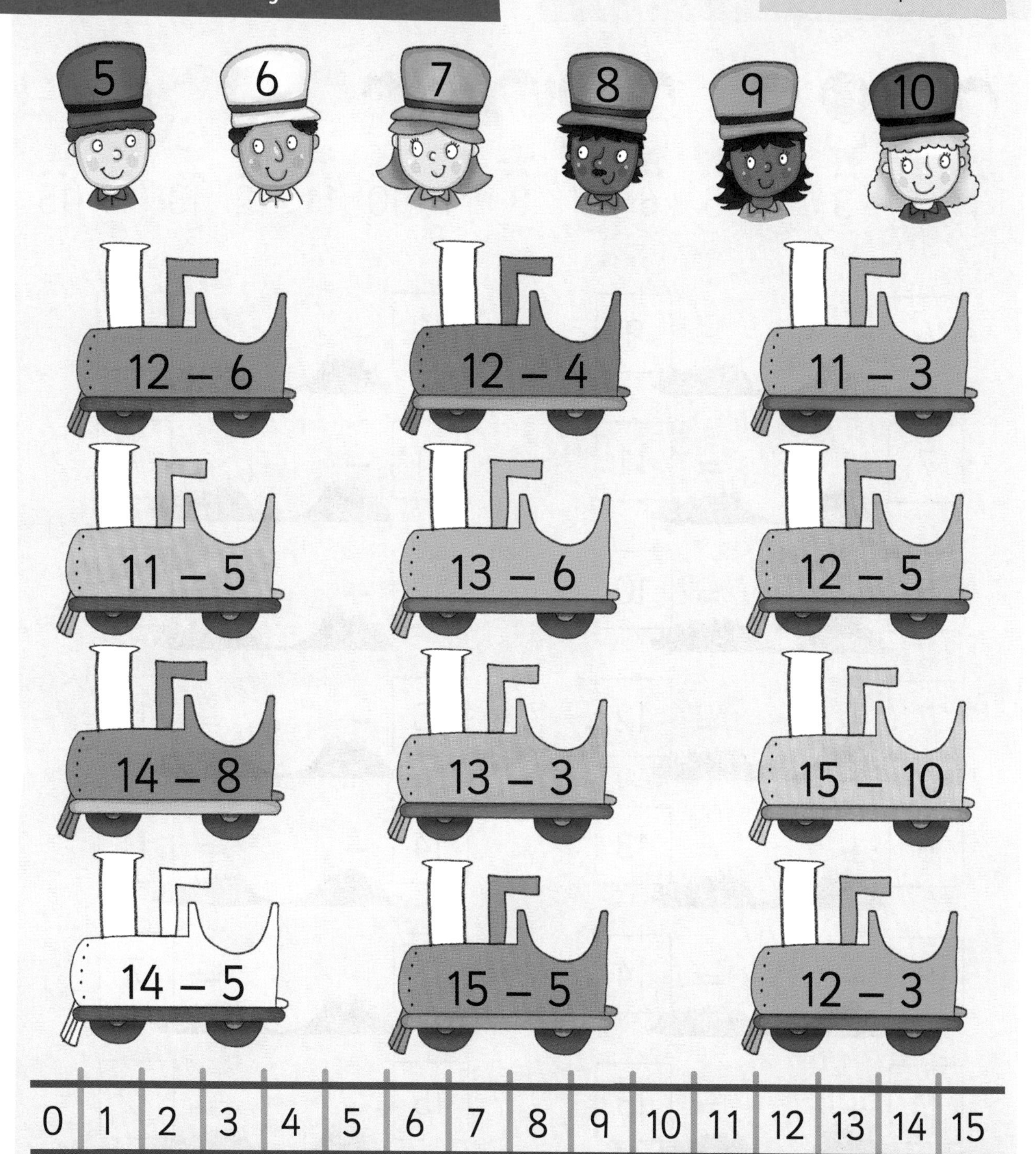

Teacher's notes

Children work out the answer to the subtraction fact on each train. They can use the number track to help them. They find the answer on a driver's hat and colour the train's funnel to match.

Date: ____________________

15 footballers

Find the missing number

$6 + \square = 9$	$12 - \square = 6$
$7 + \square = 11$	$11 - \square = 7$
$5 + \square = 10$	$12 - \square = 9$
$7 + \square = 12$	$13 - \square = 12$
$6 + \square = 13$	$14 - \square = 11$
$9 + \square = 14$	$15 - \square = 7$
$11 + \square = 15$	$15 - \square = 12$

Teacher's notes

Children work out the missing number in each of the addition and subtraction calculations and write it on the football. They can use the row of footballers to help them.

Date: ______________

Word problems

Solve problems by adding or subtracting

Samira has 12 sweets.

She gives 3 to Sam.

How many does she have left?

☐ ☐ ☐ ☐ ☐

Amber had 7 cars.

She bought 5 more.

How many cars does she have altogether?

☐ ☐ ☐ ☐ ☐

12 birds sat in a tree.

2 flew away.

How many birds were left in the tree?

☐ ☐ ☐ ☐ ☐

Hassan had 15 buttons on his shirt.

3 buttons fell off.

How many were left?

☐ ☐ ☐ ☐ ☐

Laura had 10 flowers.

Gemma gave her 5 more.

How many did Laura have altogether?

☐ ☐ ☐ ☐ ☐

Cavan has 7 stickers.

Caie has 7 stickers.

How many stickers do they have altogether?

☐ ☐ ☐ ☐ ☐

Teacher's notes

Children write the addition or subtraction calculation for each problem in the spaces provided.

Date: ______________

Cut the cakes

Find half of a shape

You will need:
- coloured pencils

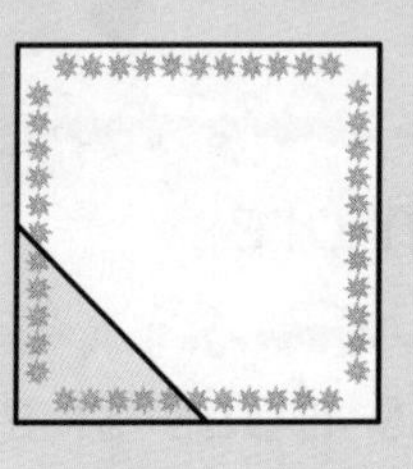
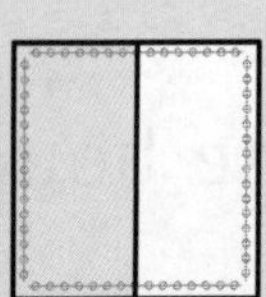
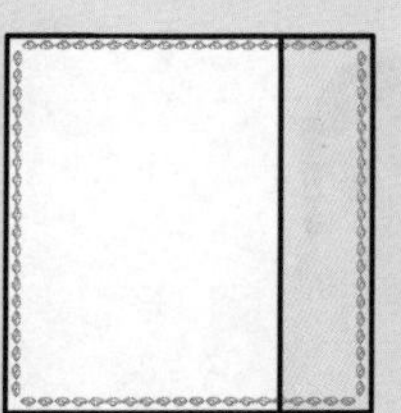
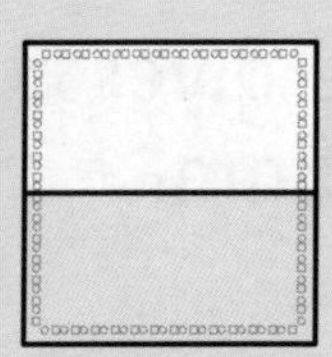

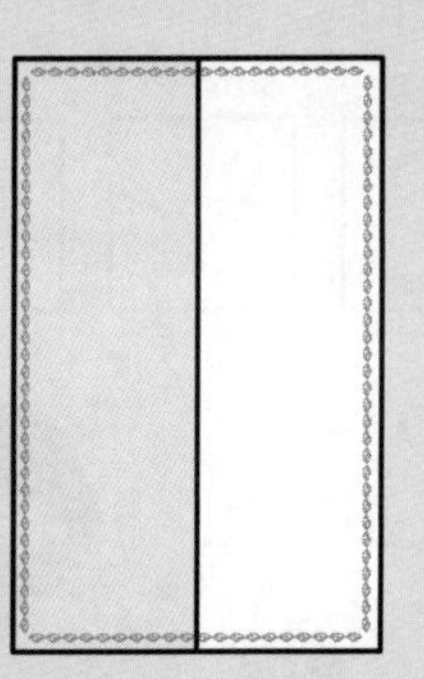
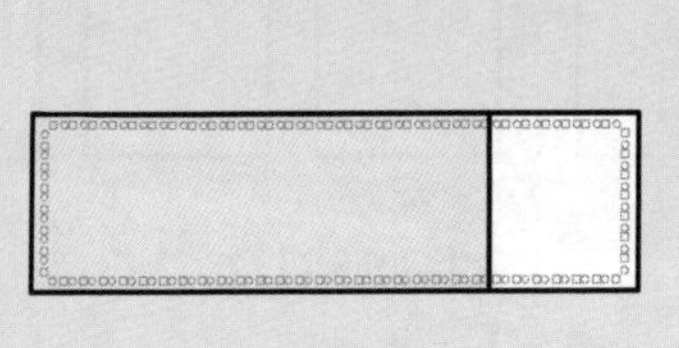
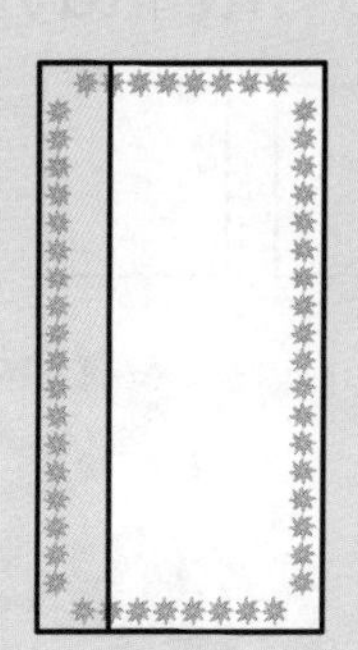
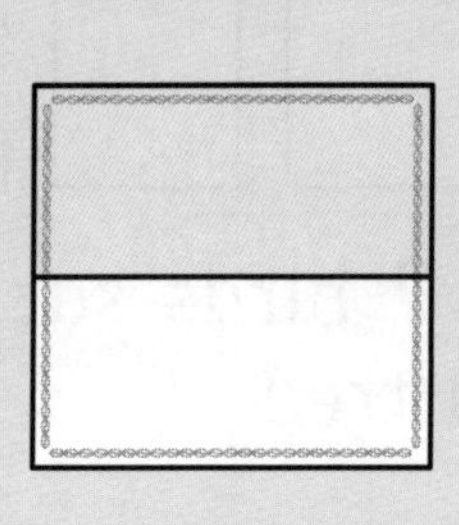

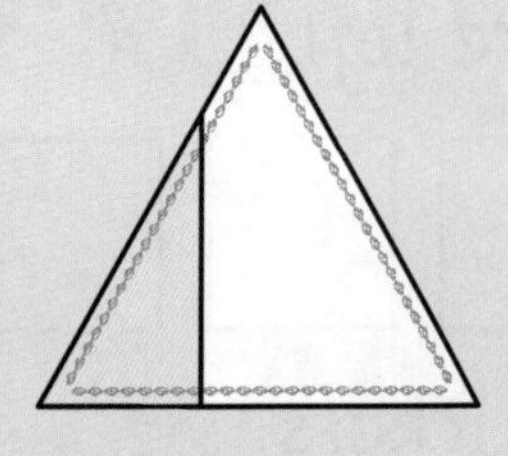
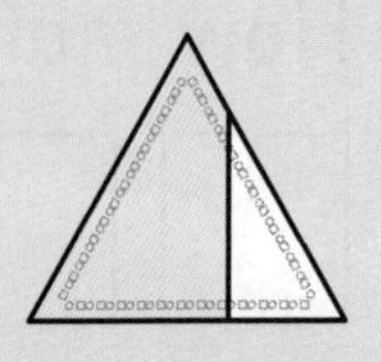
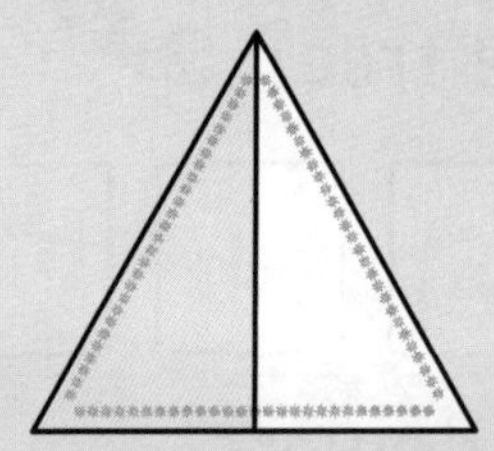
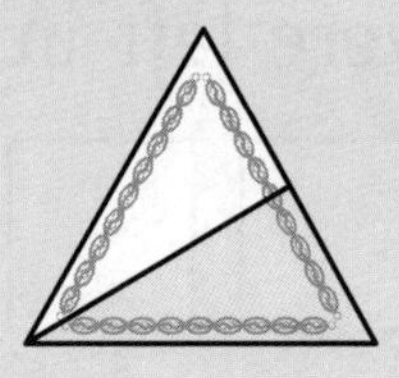

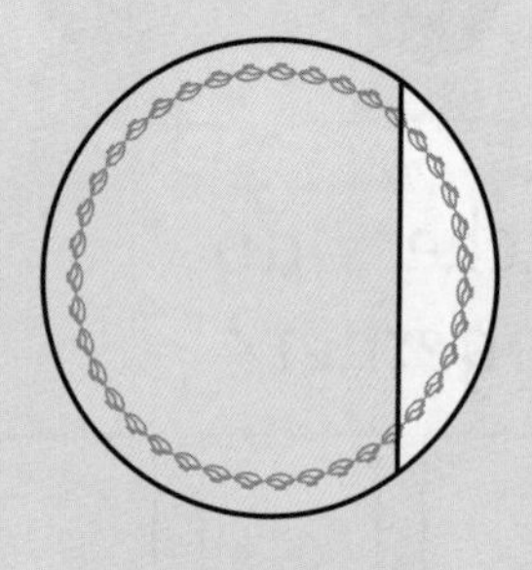

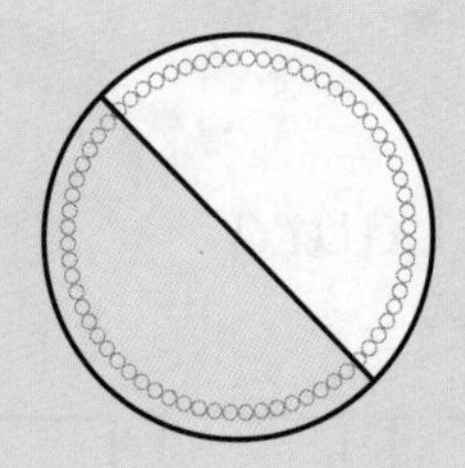
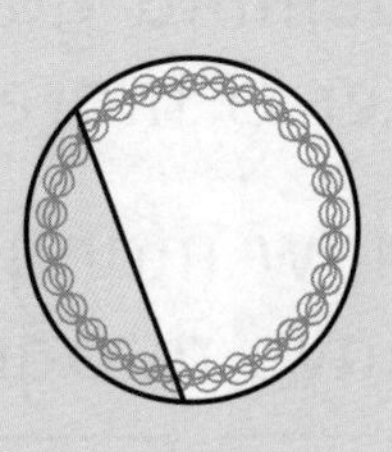

Teacher's notes

Children colour the two cakes in each row that have been cut exactly in half.

Date: ________________

Fruit tree fractions

Find half of a set of objects

Teacher's notes

For each tree, children cross out the apples and redraw them in the baskets so that they are shared equally between the two baskets. They work out half the number of apples on the tree and complete the sentence.

Date: ______________

Half a scarf

Find half of a length

You will need:
- ruler
- coloured pencils

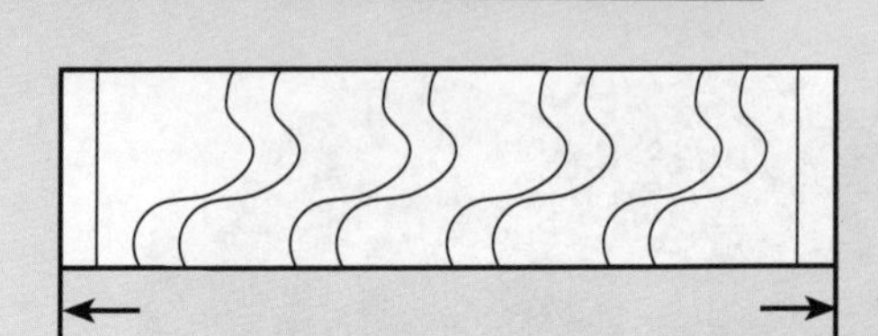

Half of this scarf measures ☐ cm.

Half of this scarf measures ☐ cm.

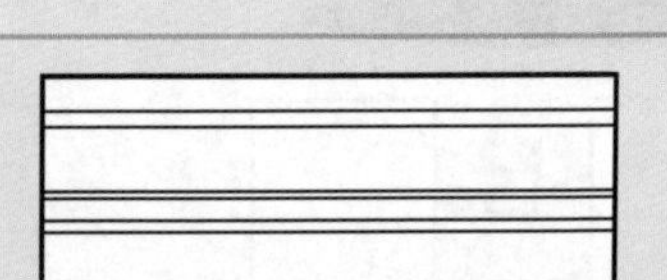

Half of this scarf measures ☐ cm.

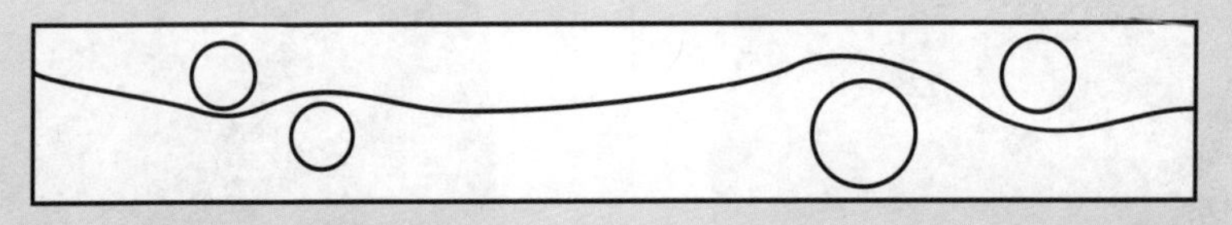

Half of this scarf measures ☐ cm.

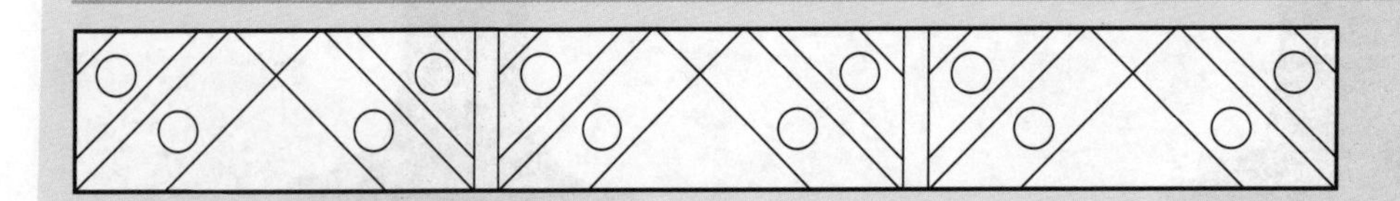

Half of this scarf measures ☐ cm.

Teacher's notes

Children use a ruler to measure the length of each scarf. They find half of this length, write it in the space provided and colour one half of the scarf.

How many?

Date: ____________

Combine halves to make one whole

You will need:
- coloured pencils

Teacher's notes

Children write how many cookie halves there are on each plate, then draw them as whole cookies and write the number in the space provided.

Date: ______________

Shopping fun

Use coins to pay for items

You will need:
- some 1p, 2p, 5p and 10p coins

Teacher's notes

Children look at the coin needed to pay for each item. They find another way of paying for the item using different coins and draw the coins in the box.

Date: ______________

Finding the same value

Show different ways of making the same value

You will need:
- some 1p and 10p coins

Teacher's notes

For the first three purses, children look at the coins and draw the number of 1p coins needed to make the same amount. For the last two purses, children look at the coins and draw the number of 10p coins needed to make the same amount.

Value of coins

Date: ____________

Recognise and understand the value of different coins

You will need:
- some 1p, 2p, 5p, 10p, 20p, 50p and £1 coins

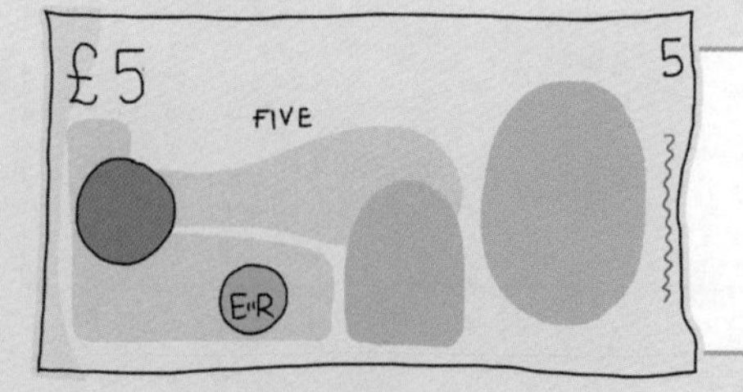

Teacher's notes

Children look at the picture of each coin or note and draw coins in the right-hand column to show the equivalent value using different coins.

Solving money problems

Solve problems about money

Date: ______________

You will need:
- some 1p, 2p and 5p coins

Items	Cost of items	Change from 10p

Items	Cost of items	Change from 10p

Teacher's notes

Children work out the cost of the items shown and write it in the table. Then they write how much change they would get from 10p.

Maths facts

Number and place value

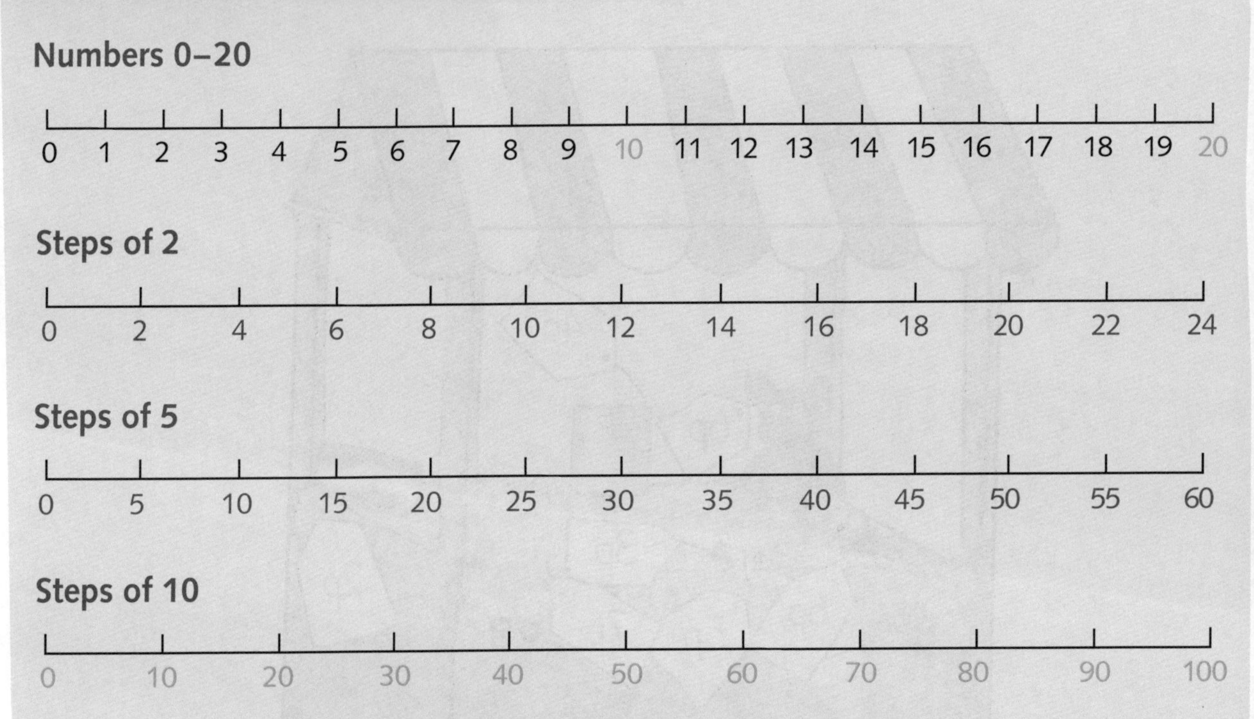

Numbers 0–20

0 1 2 3 4 5 6 7 8 9 10 11 12 13 14 15 16 17 18 19 20

Steps of 2

0 2 4 6 8 10 12 14 16 18 20 22 24

Steps of 5

0 5 10 15 20 25 30 35 40 45 50 55 60

Steps of 10

0 10 20 30 40 50 60 70 80 90 100

1–100 number square

1	2	3	4	5	6	7	8	9	10
11	12	13	14	15	16	17	18	19	20
21	22	23	24	25	26	27	28	29	30
31	32	33	34	35	36	37	38	39	40
41	42	43	44	45	46	47	48	49	50
51	52	53	54	55	56	57	58	59	60
61	62	63	64	65	66	67	68	69	70
71	72	73	74	75	76	77	78	79	80
81	82	83	84	85	86	87	88	89	90
91	92	93	94	95	96	97	98	99	100

Number facts

+	0	1	2	3	4	5	6	7	8	9	10
0	0	1	2	3	4	5	6	7	8	9	10
1	1	2	3	4	5	6	7	8	9	10	11
2	2	3	4	5	6	7	8	9	10	11	12
3	3	4	5	6	7	8	9	10	11	12	13
4	4	5	6	7	8	9	10	11	12	13	14
5	5	6	7	8	9	10	11	12	13	14	15
6	6	7	8	9	10	11	12	13	14	15	16
7	7	8	9	10	11	12	13	14	15	16	17
8	8	9	10	11	12	13	14	15	16	17	18
9	9	10	11	12	13	14	15	16	17	18	19
10	10	11	12	13	14	15	16	17	18	19	20

+	11	12	13	14	15	16	17	18	19	20
0	11	12	13	14	15	16	17	18	19	20
1	12	13	14	15	16	17	18	19	20	
2	13	14	15	16	17	18	19	20		
3	14	15	16	17	18	19	20			
4	15	16	17	18	19	20				
5	16	17	18	19	20					
6	17	18	19	20						
7	18	19	20							
8	19	20								
9	20									

Measurement (time)

4 o'clock

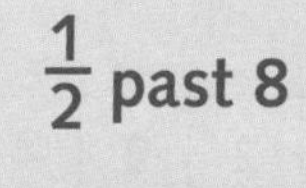

$\frac{1}{2}$ past 8

Fractions

Half: $\frac{1}{2}$

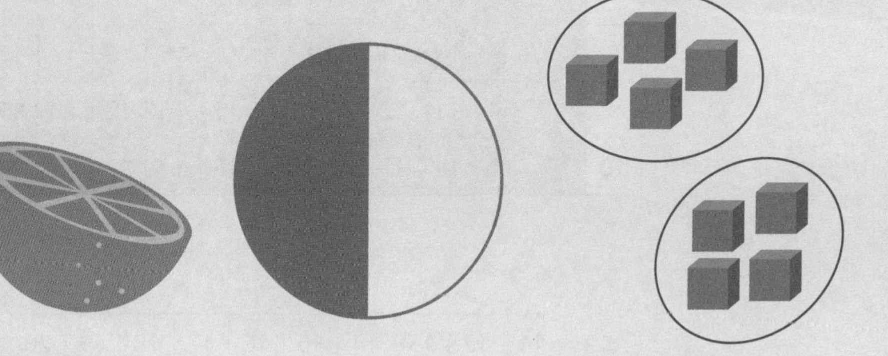

Quarter: $\frac{1}{4}$

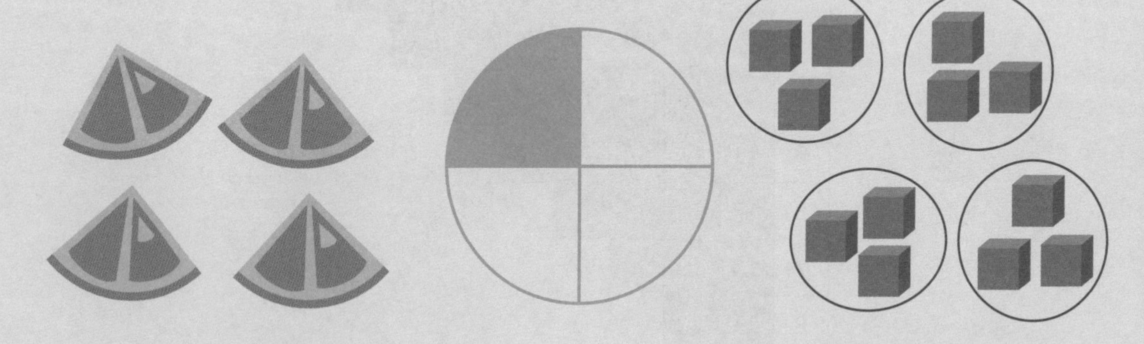

Properties of shape

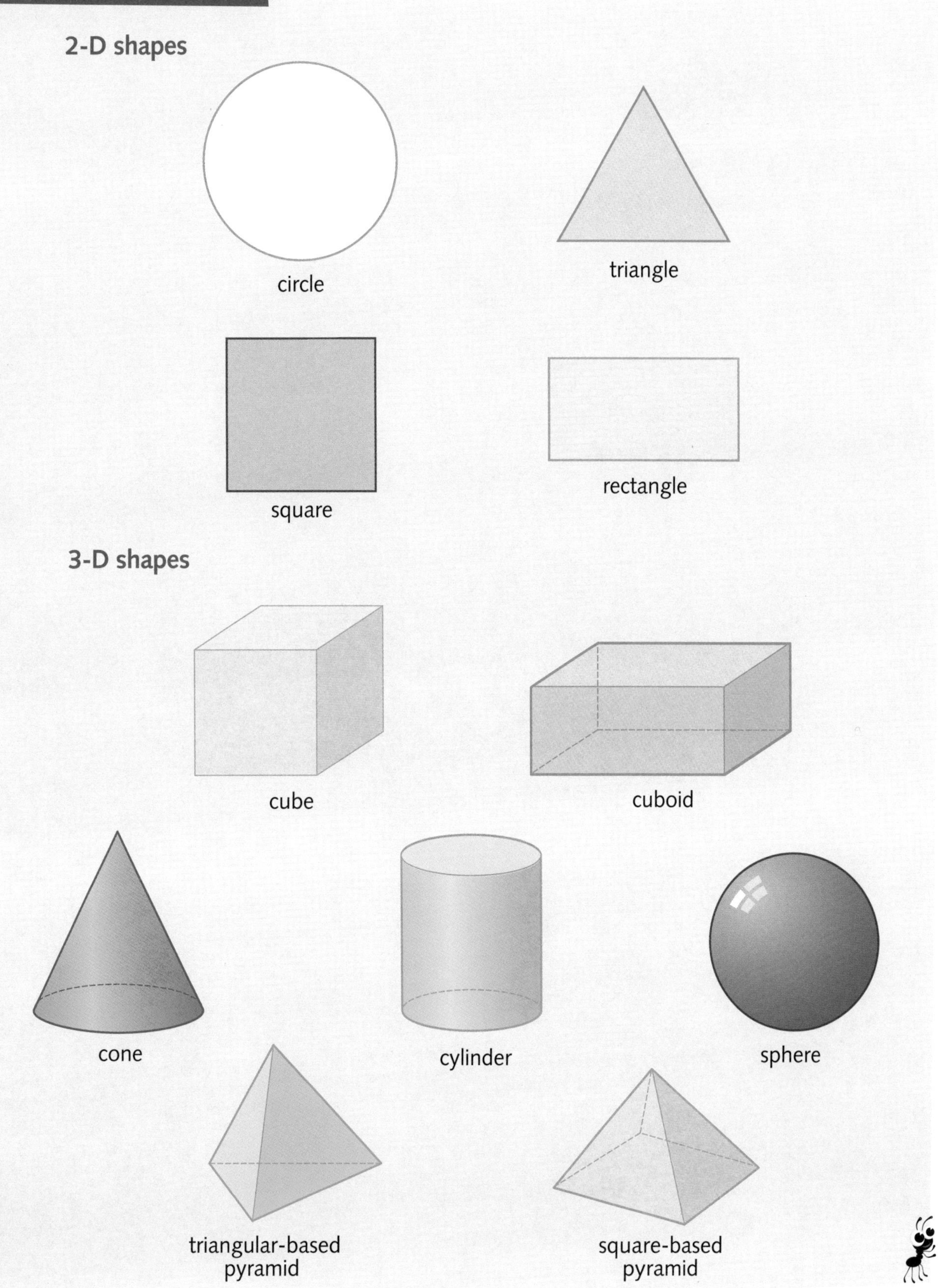

Collins

William Collins' dream of knowledge for all began with the publication of his first book in 1819.

A self-educated mill worker, he not only enriched millions of lives, but also founded a flourishing publishing house. Today, staying true to this spirit, Collins books are packed with inspiration, innovation and practical expertise. They place you at the centre of a world of possibility and give you exactly what you need to explore it.

Collins. Freedom to teach.

Published by Collins

An imprint of HarperCollins*Publishers*
77–85 Fulham Palace Road
Hammersmith
London
W6 8JB

Browse the complete Collins catalogue at
www.collins.co.uk

10 9 8 7 6 5 4 3 2
ISBN 978-0-00-756819-2

British Library Cataloguing in Publication Data

A Catalogue record for this publication is available from the British Library

Cover design and artwork: Amparo Barrera
Internal design concept: Amparo Barrera
Designers: GreenGate Publishing
Illustrators: Helen Poole, Natalia Moore, Helen Graper and Aptara

Printed by L.E.G.O

ISBN 978-0-00-756819-2